Terrific TOYS

From Around the World

By Noah Leatherland

BookLife PUBLISHING

©2023
BookLife Publishing Ltd.
King's Lynn, Norfolk
PE30 4LS, UK

All rights reserved.
Printed in China.

A catalogue record for this book is available from the British Library.

PB ISBN: 978-1-80505-123-7

Written by:
Noah Leatherland

Edited by:
Rebecca Phillips-Bartlett

Designed by:
Amelia Harris

FSC MIX Paper from responsible sources FSC® C113515

All facts, statistics, web addresses and URLs in this book were verified as valid and accurate at time of writing. No responsibility for any changes to external websites or references can be accepted by either the author or publisher.

Image Credits

All images are courtesy of Shutterstock.com, unless otherwise specified. With thanks to Getty Images, Thinkstock Photo and iStockphoto. Cover – Mauro Rodrigues, anaken2012, Alfa Photostudio, Noel V. Baebler. Used on all pages – Shmelkova Nataliya. 4 – New Africa. 5 – StockImageFactory.com, engagestock. 6&7 – Pamela Toledo. 8 – Kanizuki, DigitalStock. 9 – Mirko Kuzmanovic, Serhii Ivashchuk. 10&11 – Ezume Images. 11 – Eric Valenne geostory. 12 – Anne Marlene Arkwright, duckeesue. 13 – Dani Simmonds, claire norman. 14 – fpdress. 14&15 – Rizvisual. 16 – Lucian Coman. 17 – Anke van Wyk. 18&19 – Manu Verdier. 19 – Serge73. 20 – Rawpixel.com, Sergey Novikov. 21 – Twinsterphoto. 22 – kai keisuke, Lost in the past, Elzbieta Sekowska. 23 – EQRoy, Cris_mh.

Contents

Page 4   A World of Toys
Page 6   Chile
Page 8   China
Page 10 Ghana
Page 12 Guatemala
Page 14 Japan
Page 16 Malawi
Page 18 Mexico
Page 20 Why Are Our Toys Different?
Page 22 Fun Facts
Page 24 Glossary and Index

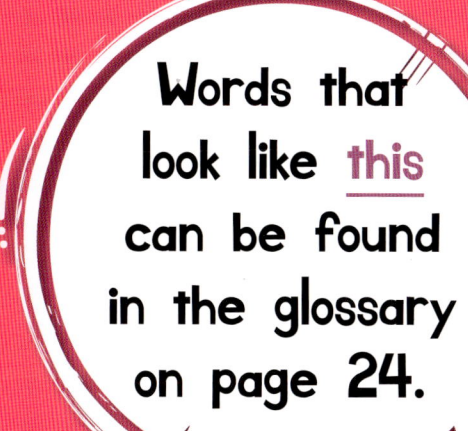

Words that look like this can be found in the glossary on page 24.

A World of Toys

From rapid race cars to cuddly teddy bears, people have come up with all sorts of toys. All around the world, there are millions of terrific toys to play with. Think of all the fun!

What is your favourite toy?

The toys you play with might be different to what children in other parts of the world play with. Some toys are **traditional** in certain places and have been around for many years.

Chile

In Chile, there is a game called emboque. It is a traditional ball and cup game. It uses a wooden stick and cap joined together with a piece of string.

Stick

Cap

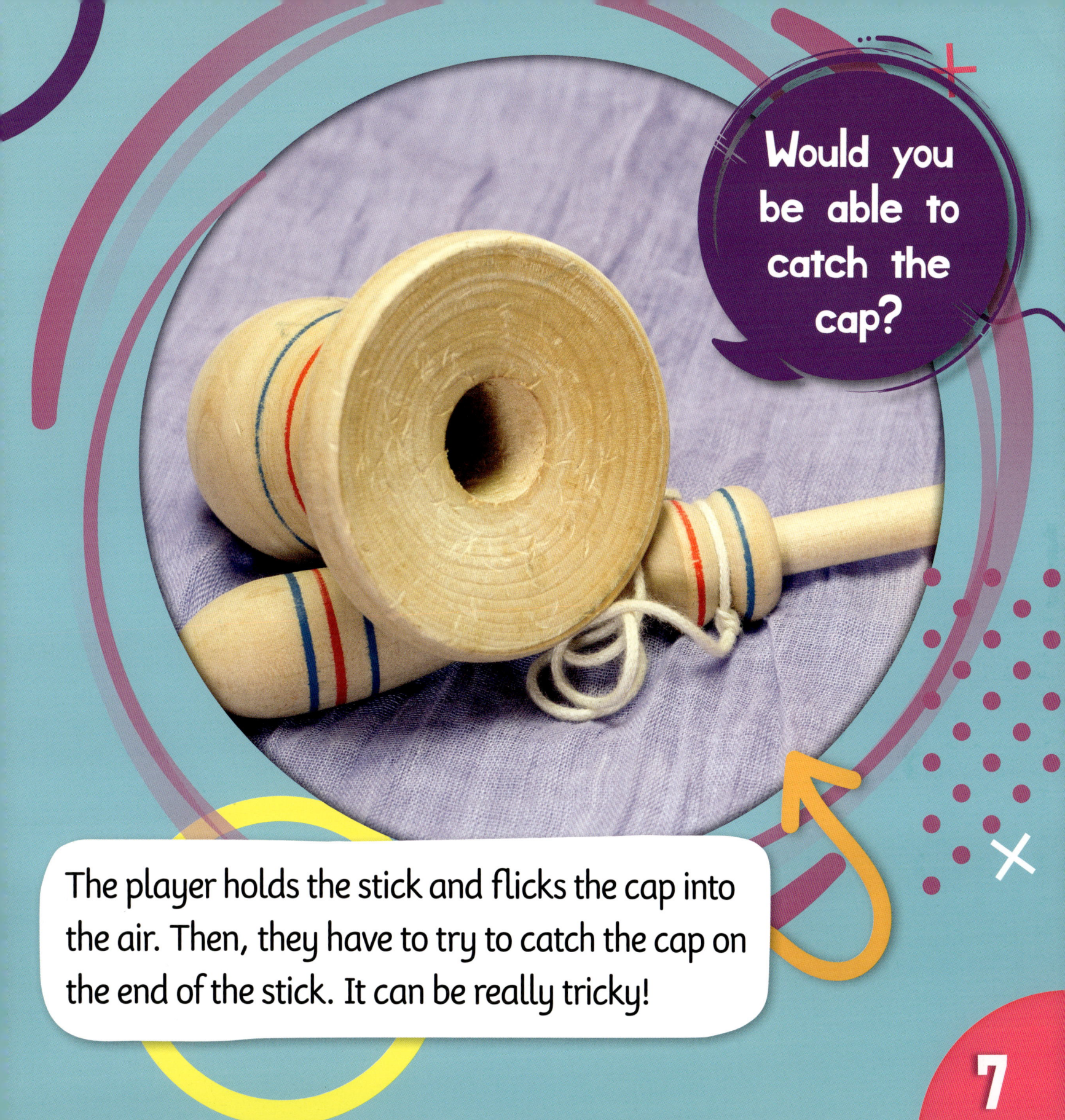

Would you be able to catch the cap?

The player holds the stick and flicks the cap into the air. Then, they have to try to catch the cap on the end of the stick. It can be really tricky!

China

In China, lots of people play with a toy called a jianzi. A jianzi is a rubber top with feathers attached, a bit like a shuttlecock. The feathers can be all sorts of colours.

Shuttlecock

The goal is to kick the jianzi and keep it up in the air.

Players can play by themselves or find another player and kick it over a net.

Ghana

Oware is a board game found all over Africa. There are lots of different <u>varieties</u> of the game. Oware is thought to have first been made in Ghana. The board might look different in other countries.

Oware is played by two people. Players try to win the game by collecting the most seeds on the board. The game is also used to help children learn to count.

The seeds can be made of stone or glass. Sometimes players use real seeds!

Guatemala

Guatemalan worry dolls are very small. Sometimes, they are only around three centimetres tall.

They are made of wood or wire and fabric. Worry dolls are dressed in traditional Guatemalan clothes.

Worry dolls are said to help people with their worries.

People say that if you tell the doll what you are worried about and put it under your pillow, you will feel much better in the morning.

Japan

Daruma otoshi is a traditional Japanese wooden toy. It is made of a **stack** of wooden blocks and a small hammer.

The top wooden block has a face painted on it.

Players use the hammer to hit the pieces out of the stack without the rest falling over. The goal is to have just the face left standing.

Do not hit the block in the face!

Malawi

In Malawi, lots of children play with galimotos. 'Galimoto' is the word for 'car' in the Chichewa language. Galimoto toys have sticks to push them around with.

Galimotos are made from recycled objects. They are often made from metal and wire. Most of the time, children make galimoto toys themselves.

Have you ever made your own toys?

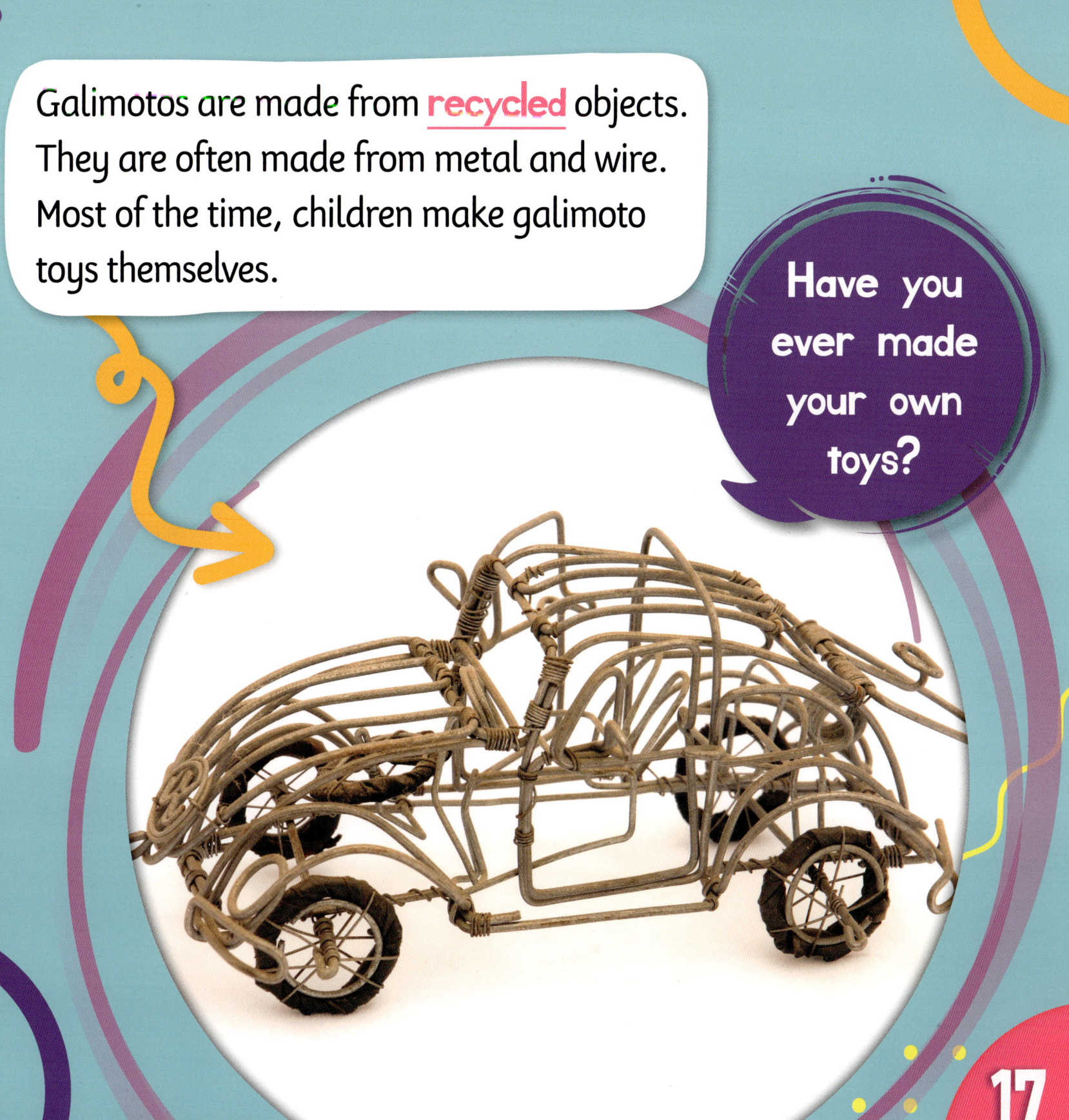

Mexico

Maria dolls are soft toys made in Mexico. The dolls have braided hair with ribbons. Their dresses are made from colourful fabrics. They are often handmade by street vendors.

Some people believe that Maria dolls help protect children from evil spirits.

Maria dolls are often given as gifts to keep people safe.

Why Are Our Toys Different?

Toys are not the only things that are different all over the world.

Different places have their own foods, clothes, languages and traditions.

Toys are an important part of each country's culture. Looking at terrific toys from around the world can teach us about different people and their cultures.

What toys from around the world would you want to play with?

Fun Facts

Spinning top toys have been found in Egyptian *tombs* from thousands of years ago.

Some different cultures end up making similar toys without realising.

A doll from Bulgaria

A doll from Eswatini

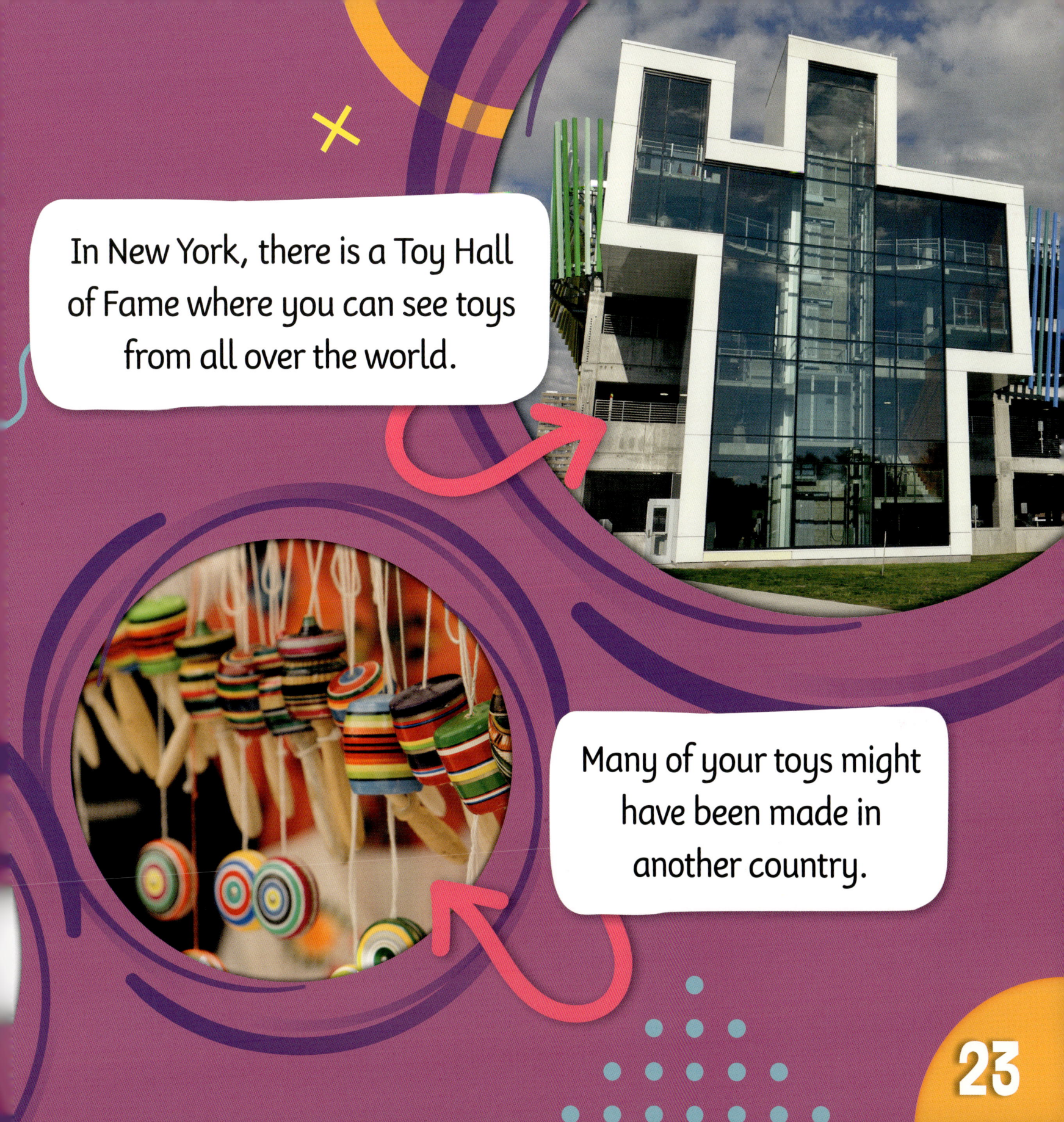

In New York, there is a Toy Hall of Fame where you can see toys from all over the world.

Many of your toys might have been made in another country.

Glossary

culture	the traditions, ideas and ways of life of a group of people
recycled	used again to make something else
spirits	beings that are not part of this world, such as a ghost or devil
stack	a pile of objects
tombs	large structures or rooms for burying the dead
traditional	something that is passed down between people over time
varieties	different types of similar things
vendors	people who sell things

Index

balls 6
blocks 14–15
cars 4, 16
dolls 12–13, 18–19, 22
games 6, 10–11
metal 17
seeds 11
wood 6, 12, 14